Commercial Pilot Practical Test Standards for Lighter-Than-Air, Balloon and Airship

NOTE

Material in FAA-S-8081-18 will be effective May 1, 1997. All previous editions of the Commercial Pilot – Lighter-Than-Air (Balloon and Airship) Practical Test Standards will be obsolete as of this date.

FOREWORD

The Commercial Pilot — Lighter-Than-Air Practical Test Standards (PTS) book has been published by the Federal Aviation Administration (FAA) to establish the standards for commercial pilot certification practical tests for the lighter-than-air category, balloon and airship classes. FAA inspectors and designated pilot examiners shall conduct practical tests in compliance with these standards. Both commercial pilots exercising flight instructor privileges and applicants should find these standards helpful during training and when preparing for the practical test.

Thomas C. Accardi

Director, Flight Standards Service

CONTENTS

INTRODUCTION

General Information

The Flight Standards Service of the Federal Aviation Administration (FAA) has developed this practical test book as the standard to be used by FAA inspectors and designated pilot examiners when conducting commercial pilot — lighter-than-air (balloon and airship) practical tests. Commercial lighter-than-air pilots exercising instructor privileges are expected to use this book when preparing applicants for practical tests. Applicants should be familiar with this book and refer to these standards during their training.

Information considered directive in nature is described in this practical test book in terms such as "shall" and "must" indicating the actions are mandatory. Guidance information is described in terms such as "should" and "may" indicating the actions are desirable or permissive but not mandatory.

The FAA gratefully acknowledges the valuable assistance provided by many individuals and companies who contributed their time and talent in assisting with the revision of these practical test standards.

Practical Test Standard Concept

Title 14 of the Code of Federal Regulations (14 CFR) part 61 specifies the areas in which knowledge and skill shall be demonstrated by an applicant before issuance of a commercial pilot certificate with the associated category and class ratings. This regulation provides the flexibility that permits the FAA to publish practical test standards containing specific TASKS in which competency shall be demonstrated. The FAA will revise this book whenever it is determined that changes are needed in the interest of safety. Adherence to the provisions of the regulations and the practical test standards is mandatory for evaluation of commercial pilot applicants.

Practical Test Book Description

This book contains the practical test standards for commercial pilot — lighter-than-air

Section 1	Lighter-Than-Air, Balloon
Section 2	Lighter-Than-Air, Airship

The Commercial Pilot — Lighter-Than-Air Practical Test Standards include the AREAS OF OPERATION and TASKS for the issuance of an initial commercial pilot certificate and for the addition of category and/or class ratings to that certificate.

Practical Test Standard Description

AREAS OF OPERATION are phases of the practical test arranged in a logical sequence within each standard. They begin with Fundamentals of Instructing and end with Postflight Procedures. TASKS are knowledge areas, flight procedures, or maneuvers appropriate to an AREA OF OPERATION. However, the examiner may conduct the practical test in any sequence that results in a complete and efficient test.

In Section I, an abbreviation within parentheses immediately following a TASK title refers to the type of balloon to which that TASK applies. Absence of an abbreviation indicates the TASK is appropriate for all balloons.

LBG	Lighter-Than-Air, Balloon (Gas)
LBH	Lighter-Than-Air, Balloon (with Airborne Heater)

NOTE is used to emphasize special considerations required in the AREA OF OPERATION or TASK.

REFERENCE identifies the publication(s) that describes the TASK. Descriptions of TASKS are not included in the standards because this information can be found in the REFERENCES listed for each TASK. Publications other than those listed may be used as references if their content conveys substantially the same meaning as the referenced publication. REFERENCES listed in this book include the current revisions of the following publications:

14 CFR part 1	Definitions and Abbreviations
14 CFR part 43	Maintenance, Preventative Maintenance, Rebuilding, and Alteration
14 CFR part 61	Certification: Pilots and Flight Instructors
14 CFR part 91	General Operating and Flight Rules
NTSB 830	Notification and Reporting of Aircraft Accidents and Incidents
AC 00-2	Advisory Circular Checklist
AC 00-6	Aviation Weather
AC 00-45	Aviation Weather Services
AC 60-14	Aviation Instructor's Handbook
AC 61-21	Flight Training Handbook
AC 61-23	Pilot's Handbook of Aeronautical Knowledge
AC 61-65	Certification: Pilots and Flight Instructors
AC 61-67	Stall Spin Awareness Training

AC 61-84	Role of Preflight Preparation
AC 61-98	Currency and Additional Qualification Requirements for Certificated Pilots
AC 61-101	Presolo Written Test
AC 90-48	Pilots' Role in Collision Avoidance
AC 91-13	Cold Weather Operation of Aircraft
AC 91-71	Operation of Hot Air Balloons with Airborne Heaters
A/FD	Airport/Facility Directory
AIM	Aeronautical Information Manual
NOTAM's	Notices to Airmen
Other	Balloon Digest (Balloon Federation of America) How To Fly A Balloon (Balloon Publishing Co.) Aerostatics (US Air Corps) Balloon and Airship Flight Manuals Navigation Equipment Operations Manuals Selected Navigation Charts

Each TASK has an Objective consisting of a series of elements. The examiner determines that the applicant meets the TASK Objective through the demonstration of competency in various elements of knowledge and/or skill. The Objectives of TASKS in certain AREAS OF OPERATION, such as Fundamentals of Instructing and Technical Subjects, include only knowledge elements.

Use of the Practical Test Standards Book

The FAA requires that all practical tests be conducted in accordance with the appropriate Commercial Pilot Practical Test Standard and the policies set forth in the INTRODUCTION. Commercial pilot applicants shall be evaluated in all TASKS included in the AREAS OF OPERATION of the appropriate practical test standard.

In preparation for each practical test, the examiner shall develop a written "plan of action." The "plan of action" shall include all TASKS (unless noted otherwise) in each AREA OF OPERATION. The "plan of action" for a practical test for the addition of an aircraft category and/or class rating shall include the required AREAS OF OPERATION and TASKS as indicated in the table at the beginning of each standard. Any TASK selected for evaluation during a practical test shall be evaluated in its entirety.

The examiner is not required to follow the precise order in which AREAS OF OPERATION and TASKS appear in this book. The examiner may change the sequence or combine TASKS with similar Objectives to have an orderly and efficient flow of the practical test. However, the Objectives of all TASKS shall be demonstrated and evaluated during the practical test.

In AREAS OF OPERATION I, II, and IV, the term "instructional knowledge" means the "what," "why," and "how" of a subject matter topic, procedure, or maneuver. It also means that the applicant's discussions, explanations, and descriptions should follow the recommended teaching procedures and techniques explained in AC 60-14, Aviation Instructor's Handbook.

The examiner shall not simulate any condition that may jeopardize safe flight or result in possible damage to the aircraft.

The examiner shall place special emphasis upon areas of aircraft operation that are most critical to flight safety. Among these are precise aircraft control and sound judgment in decision making. Although these areas may or may not be listed under each TASK, they are essential to flight safety and should receive careful evaluation throughout the practical test. THE EXAMINER SHALL ALSO EMPHASIZE COLLISION AVOIDANCE, CHECKLIST USAGE, AND USE OF DISTRACTIONS.

Commercial Pilot — Lighter-Than-Air Practical Test Prerequisites

An applicant for an initial commercial pilot certification practical test is required by 14 CFR part 61 to:

1. have reached the age of 18 years;
2. hold at least a private pilot certificate;[1]
3. have the prescribed aeronautical experience and training required for the category and class rating sought;
4. have passed the appropriate commercial pilot knowledge test since the beginning of the 24th month before the month in which the applicant takes the practical test; and
5. obtain an endorsement from an authorized instructor certifying that the applicant has been given flight instruction in the AREAS OF OPERATION listed in 14 CFR part 61, section 61.127, for the commercial pilot rating sought in preparation for the practical test within 60 days preceding the date of application. The statement shall also state that the instructor finds the applicant competent to pass the practical test, and that the applicant has satisfactory knowledge of the subject area(s) in which a deficiency was indicated on the knowledge test report.2

An applicant already holding a commercial pilot certificate who applies for an additional category and/or class rating on that certificate must:

1 This requirement is based on a proposed revision to 14 CFR part 61.

2 AC 61-65, Certification: Pilots and Flight Instructors, states that the instructor may sign the recommendation on the reverse side of FAA Form 8710-1, Airman Certificate and/or Rating Application, in lieu of the previous statement, provided all appropriate 14 CFR part 61 requirements are substantiated by reliable records.

1. hold a commercial pilot certificate;
2. have the prescribed aeronautical experience and training required for the category and class rating sought;
3. pass an additional knowledge test for the rating sought (unless already rated in airplane, rotorcraft, or airship) since the beginning of the 24th month before the month in which he/she takes the practical test; and
4. obtain an endorsement from an authorized instructor certifying that the applicant has been given flight instruction in the applicable AREAS OF OPERATION listed in 14 CFR part 61, section 61.127, for the commercial pilot rating sought in preparation for the practical test within 60 days preceding the date of application. The statement shall also state that the instructor finds the applicant competent to pass the practical test, and that the applicant has satisfactory knowledge of the subject area(s) in which a deficiency was indicated on the knowledge test report2.

Aircraft and Equipment Required for the Practical Test

The applicant is required by 14 CFR part 61, section 61.45, to provide an airworthy, certificated aircraft for use during the practical test. This section further requires that the aircraft be capable of performing all appropriate TASKS for the commercial pilot rating sought and have no operating limitations which prohibit the performance of those operations.

Examiner[3] Responsibility

1. The examiner conducting the practical test is responsible for determining that the applicant meets the acceptable standards of knowledge and skill of each TASK within the appropriate practical test standard. The examiner makes this determination by accomplishing the Objective for each selected TASK, and includes an evaluation of the applicant'sability to apply fundamentals of instructing, when required;
2. knowledge of and ability to teach subject matter covered in required TASKS; and
3. ability to perform procedures and maneuvers included in the standards to the COMMERCIAL PILOT skill level.

Although it is the examiner's primary concern to observe the applicant's ability to skillfully and safely operate the aircraft, oral questioning may be used at any time during the practical test to determine that the applicant has a comprehensive knowledge of the TASKS and their related safety factors. During the flight portion of the practical test, the examiner shall evaluate use of visual scanning and collision avoidance procedures.

3 The word "examiner" is used throughout the standard to denote either the FAA inspector or FAA designated pilot examiner who conducts an official practical test.

Satisfactory Performance

1. The practical test is passed if, in the judgment of the examiner, the applicant demonstrates satisfactory performance with regard to: knowledge of fundamentals of instructing;
2. knowledge of technical subject areas;
3. knowledge of flight instructor responsibilities concerning the pilot certification process;
4. knowledge of flight instructor responsibilities concerning log book entries and pilot certificate endorsements;
5. knowledge of development and effective use of a course of training, a syllabus, and a lesson plan; and
6. ability to demonstrate all required procedures and maneuvers to the COMMERCIAL PILOT skill level.

Unsatisfactory Performance

1. If, in the judgment of the examiner, the applicant does not meet the standards of performance of any TASK performed, the associated AREA OF OPERATION is considered unsatisfactory and, therefore, the practical test is failed. The examiner or applicant may discontinue the test at any time when failure of an AREA OF OPERATION makes the applicant ineligible for the certificate or rating sought. The test may be continued only with the consent of the applicant. If the test is discontinued, the applicant is entitled to credit for only those AREAS OF OPERATION satisfactorily performed. However, during the retest and at the discretion of the examiner, any TASK may be re-evaluated, including those previously passed. Specific reasons for disqualification are: failure to perform a procedure or maneuver to the COMMERCIAL PILOT skill level;
2. any action or lack of action by the applicant which requires corrective intervention by the examiner to maintain safe flight; and
3. failure to use proper and effective visual scanning techniques to clear the area before and while performing maneuvers.

When a disapproval notice is issued, the examiner shall record the applicant's unsatisfactory performance in terms of AREAS OF OPERATION appropriate to the practical test conducted.

Crew Resource Management (CRM)

CRM refers to the effective use of all available resources including human, hardware, and information. Human resources includes all groups routinely working with the pilot (or cockpit crew) who are involved in decisions required to operate a flight safely. These groups include, ground crew, chase crew, maintenance personnel, and weather services. CRM is not a single TASK but a set of knowledge and skill competencies that must be evident in all TASKS in this practical test standard as applied to either single pilot or a crew operation.

Applicant's Use of Checklists

Throughout the practical test, the applicant is evaluated on the use of an appropriate checklist. Proper use is dependent on the specific TASK being evaluated. The situation may be such that the use of the checklist, while accomplishing elements of an Objective, would be either unsafe or impractical, especially in a single-pilot operation. In this case, a review of the checklist after the elements have been accomplished would be considered appropriate. Use of a checklist should also take into consideration visual scanning and division of attention at all times.

Use of Distractions During Practical Tests

Numerous studies indicate that many accidents have occurred when the pilot has been distracted during critical phases of flight. To evaluate the pilot's ability to utilize proper control technique while dividing attention both inside and/or outside the basket or car, the examiner shall cause a realistic distraction during the flight portion of the practical test to evaluate the applicant's ability to divide attention while maintaining safe flight.

Metric Conversion Initiative

To assist the pilots in understanding and using the metric measurement system, the practical test standards refer to the metric equivalent of various altitudes throughout. The inclusion of meters is intended to familiarize pilots with its use. The metric altimeter is arranged in 10 meter increments; therefore, when converting from feet to meters, the exact conversion, being too exact for practical purposes, is rounded to the nearest 10 meter increment or even altitude as necessary.

Flight Instructor Responsibility

A commercial lighter-than-air pilot, when exercising flight instructor privileges, is responsible for training pilot applicants to acceptable standards in all TASKS appropriate for the rating sought. Because of the impact of their teaching activities in developing safe, proficient pilots, instructors should exhibit a high level of knowledge, skill, and the ability to impart that knowledge and skill to students. The instructor shall certify that the applicant

1. able to exhibit a practical application of the fundamentals of instructing;
2. competent to teach the subject matter, procedures, and maneuvers included in the standards to students with varying backgrounds and levels of experience and ability;

3. able to perform the procedures and maneuvers included in the standards to the COMMERCIAL PILOT skill level; and
4. competent to pass the required practical test for issuance of the commercial pilot certificate with the associated category and class ratings, or the addition of a category and/or class rating to a commercial pilot certificate.

Throughout the applicant's training, the instructor is responsible for emphasizing the performance of, and the ability to teach, effective visual scanning and collision avoidance procedures. These areas are covered in AC 90-48, Pilots' Role in Collision Avoidance; AC 61-21, Flight Training Handbook; AC 61-23, Pilot's Handbook of Aeronautical Knowledge; and the Aeronautical Information Manual.

Instrument Privileges — Airship

14 CFR part 61, section 61.5, states that no person may act as pilot in command of a civil aircraft under instrument flight rules, or in weather conditions less than the minimums prescribed for VFR flight unless, in the case of an airship, they hold a commercial pilot certificate with lighter-than-air category and airship class ratings.

In order to allow an efficient and economical practical test for commercial pilot certification in an airship, the following considerations have been made regarding evaluation of instrument TASKS.

1. If the applicant already holds an instrument rating, instrument TASKS need not be evaluated during the practical test. However, the examiner shall verify the applicant's training records to determine that all instrument training in an airship required by regulation has been received and that the training records have been properly endorsed by the instructor giving the training.
2. If the applicant does not hold an instrument rating, the examiner shall verify the applicant's logbook to determine that the applicant has been trained in all required instrument operations in an airship required by regulation and that the training records have been properly endorsed by the instructor giving the training. During the practical test, instrument skills shall be evaluated using FAA-S-8081-4, Instrument Rating Practical Test Standards. The examiner shall select appropriate TASKS from AREA OF OPERATION IV and at least one instrument approach TASK from AREA OF OPERATION VI for evaluation during the practical test.

SECTION 1

COMMERCIAL PILOT LIGHTER-THAN-AIR — BALLOON

Practical Test Standards

CONTENTS

Lighter-Than-Air —Balloon

Addition of a Lighter-Than-Air/Balloon rating to an existing Commercial Pilot Certificate

Required TASKS are indicated by either the TASK letter(s) that apply(s) or an indication that all or none of the TASKS must be tested.						
Area of Operation	Rating(s) Held					
	ASE	AME	RH	RG	G	LA
I	ALL	ALL	ALL	ALL	ALL	NONE
II	B,C,F	B,C,F	B,C,F	B,C,F	B,C,F	B,C
III	D,E	D,E	D,E	D,E	D,E	D,E
IV	ALL	ALL	ALL	ALL	ALL	NONE
V	ALL	ALL	ALL	ALL	ALL	ALL
VI	NONE	NONE	NONE	NONE	NONE	NONE
VII	ALL	ALL	ALL	ALL	ALL	ALL
VIII	ALL	ALL	ALL	ALL	ALL	ALL
IX	NONE	NONE	NONE	NONE	NONE	NONE
X	A,C,D	A,C,D	A,C,D	A,C,D	ALL	A,C,D
XI	ALL	ALL	ALL	ALL	ALL	ALL

LEGEN

ASE	Airplane Single-Engine
AME	Airplane Multiengine
RH	Rotorcraft Helicopter
RG	Rotorcraft Gyroplane
G	Glider
LA	Lighter-Than-Air Airship

If an applicant holds more than one rating on a commercial pilot certificate and the table indicates both ALL and NONE for a particular AREA OF OPERATION, the NONE entry applies. This is logical since the applicant has satisfactorily accomplished the AREA OF OPERATION on a previous commercial pilot practical test. If an applicant holds a flight instructor certificate, AREAS OF OPERATION I and IV are not required. However, at the discretion of the examiner, the applicant's competence in all AREAS OF OPERATION may be evaluated.

APPLICANT'S PRACTICAL TEST CHECKLIST (BALLOON)

APPOINTMENT WITH EXAMINER

EXAMINER'S NAME ______________________________

LOCATION ______________________________

DATE/TIME ______________________________

ACCEPTABLE AIRCRAFT

- ○ Aircraft Documents:

  Airworthiness Certificate
  Registration Certificate
  Operating Limitations

- ○ Aircraft Maintenance Records:

  Logbook Record of Airworthiness Inspections and AD Compliance

- ○ Balloon Flight Manual

PERSONAL EQUIPMENT

- ○ Practical Test Standard
- ○ Current Aeronautical Chart
- ○ Computer and Plotter
- ○ Flight Log
- ○ Current AIM

PERSONAL RECORDS

- ○ Identification - Photo/Signature ID
- ○ Pilot Certificate
- ○ Current and Appropriate Medical Certificate or Statement
- ○ Completed FAA Form 8710-1, Airman Certificate and/or Rating Application with Instructor's Signature
- ○ AC Form 8080-2, Airman Written Test Report or Computer Test Report

- Pilot Logbook with Appropriate Instructor Endorsements
- FAA Form 8060-5, Notice of Disapproval (if applicable)
- Approved School Graduation Certificate (if applicable)
- Examiner's Fee (if applicable)

EXAMINER'S PRACTICAL TEST CHECKLIST (BALLOON)

APPLICANT'S NAME____

LOCATION____

DATE/TIME ____

I. FUNDAMENTALS OF INSTRUCTING

- ○ A. The Learning Process
- ○ B. Human Behavior
- ○ C. The Teaching Process
- ○ D. Teaching Methods
- ○ E. Critique and Evaluation
- ○ F. Flight Instructor Characteristics and Responsibilities
- ○ G. Planning Instructional Activity

II. TECHNICAL SUBJECTS

- ○ A. Aeromedical Factors
- ○ B. Visual Scanning and Collision Avoidance
- ○ C. Principles of Flight
- ○ D. Regulations and Publications
- ○ E. National Airspace System
- ○ F. Logbook Entries and Certificate Endorsements

III. PREFLIGHT PREPARATION

- ○ A. Certificates and Documents
- ○ B. Weather Information
- ○ C. Flight Planning
- ○ D. Performance and Limitations
- ○ E. Operation of Systems

IV. PREFLIGHT LESSON ON A MANEUVER TO BE PERFORMED IN FLIGHT

- Maneuver Lesson

V. PREFLIGHT PROCEDURES

- A. Launch Site Selection
- B. Crew Briefing and Preparation
- C. Layout and Assembly
- D. Preflight Inspection
- E. Inflation
- F. Basket/Gondola Management
- G. Pre-Launch Check

VI. AIRPORT OPERATIONS

- Radio Communications

VII. LAUNCHES AND LANDINGS

- A. Normal Launch
- B. Launch Over Obstacle
- C. Approach to Landing
- D. Steep Approach to Landing
- E. Normal Landing
- F. High-Wind Landing

VIII. PERFORMANCE MANEUVERS

- A. Ascents
- B. Altitude Control (Level Flight)
- C. Descents
- D. Rapid Ascent and Descent
- E. Contour Flying (LBH)

- F. High Altitude Flight (LBG)
- G. Obstacle Avoidance (LBH)
- H. Tethering (LBH)
- I. Winter Flying
- J. Mountain Flying

IX. NAVIGATION

- Navigation

X. EMERGENCY OPERATIONS

- A. Systems and Equipment Malfunctions
- B. Emergency Equipment and Survival Gear
- C. Water Landing
- D. Thermal Flight

XI. POSTFLIGHT PROCEDURES

- A. Recovery
- B. Deflation and Pack-up
- C. Refueling (LBH)

I. AREA OF OPERATION: FUNDAMENTALS OF INSTRUCTING

NOTE: The examiner will select TASK F and at least one other TASK.

A. TASK: THE LEARNING PROCESS

REFERENCE: AC 60-14.

Objective. To determine that the applicant exhibits instructional knowledge of the elements of the learning process by describing:

1. The definition and characteristics of learning.
2. Practical application of the laws of learning.
3. Factors involved in how people learn.
4. Recognition and proper use of the various levels of learning.
5. Principles that are applied in learning a skill.
6. Factors of forgetting and retention.
7. How the transfer of learning affects the learning process.
8. How the formation of habit patterns affects the learning process.

B. TASK: HUMAN BEHAVIOR

REFERENCE: AC 60-14.

Objective. To determine that the applicant exhibits instructional knowledge of the elements of human behavior by describing:

1. Control of human behavior.
2. Development of student potential.
3. Relationship of human needs to behavior and learning.
4. Relationship of defense mechanisms to student learning and pilot decision making.
5. General rules which a flight instructor should follow during student training to ensure good human relations.

C. TASK: THE TEACHING PROCESS

REFERENCE: AC 60-14.

Objective. To determine that the applicant exhibits instructional knowledge of the elements of the teaching process by describing:

1. Preparation of a lesson for a ground or flight instructional period.
2. Presentation of knowledge and skills, including the methods which are suitable in particular situations.
3. Application, by the student, of the knowledge and skills presented by the instructor.
4. Review of the material presented and the evaluation of student performance and accomplishment.

D. TASK: TEACHING METHODS

REFERENCE: AC 60-14.

Objective. To determine that the applicant exhibits instructional knowledge of the elements of teaching methods by describing:

1. The organization of a lesson, i.e., introduction, development, and conclusion.
2. The lecture method.
3. The guided discussion method.
4. The demonstration-performance method.
5. Computer/video assisted instruction.

E. TASK: CRITIQUE AND EVALUATION

REFERENCE: AC 60-14.

Objective. To determine that the applicant exhibits instructional knowledge of the elements of critique and evaluation by describing:

1. Purpose and characteristics of an effective critique.
2. Difference between critique and evaluation.
3. Characteristics of effective oral questions and what type to avoid.
4. Responses to student questions.
5. Characteristics and development of effective written tests.
6. Characteristics and uses of performance tests, specifically, the FAA practical test standards.

F. TASK: FLIGHT INSTRUCTOR CHARACTERISTICS AND RESPONSIBILITIES

REFERENCE: AC 60-14.

Objective. To determine that the applicant exhibits instructional knowledge of the elements of flight instructor characteristics and responsibilities by describing:

1. Characteristics and qualifications of a professional flight instructor.
2. Role of the flight instructor in dealing with student stress, anxiety, and psychological abnormalities.
3. Flight instructor's responsibility with regard to student pilot supervision and surveillance.
4. Flight instructor's authority and responsibility for endorsements and recommendations.
5. Flight instructor's responsibility in the conduct of the required FAA flight review.

G. TASK: PLANNING INSTRUCTIONAL ACTIVITY

REFERENCE: AC 60-14.

Objective. To determine that the applicant exhibits instructional knowledge of the elements of planning instructional activity by describing:

1. Development of a course of training.
2. Content and use of a training syllabus.
3. Purpose, characteristics, proper use, and items of a lesson plan.
4. Flexibility features of a course of training, syllabus, and lesson plan required to accommodate students with varying backgrounds, levels of experience, and ability.

II. AREA OF OPERATION: TECHNICAL SUBJECTS

A. TASK: AEROMEDICAL FACTORS

REFERENCES: AC 61-21; AIM.

Objective. To determine that the applicant exhibits instructional knowledge of the elements of aeromedical factors by describing:

1. Hypoxia, its symptoms, effects, and corrective action.
2. Hyperventilation, its symptoms, effects, and corrective action.
3. Middle ear and sinus problems, their causes, effects, and corrective action.
4. Effects of alcohol and drugs, and their relationship to safety.
5. How evolved gas from scuba diving can affect a pilot during flight.
6. Fatigue, its effects and corrective action.

B. TASK: VISUAL SCANNING AND COLLISION AVOIDANCE

REFERENCES: AC 61-21, AC 90-48; AIM.

Objective. To determine that the applicant exhibits instructional knowledge of the elements of visual scanning and collision avoidance by describing:

1. Relationship between a pilot's physical or mental condition and vision.
2. Practice of "time sharing" of attention inside and outside the basket.
3. Appropriate visual scanning techniques.
4. Importance of controlling ascents and descents.
5. Situations which involve the greatest collision risk.

C. TASK: PRINCIPLES OF FLIGHT

REFERENCES: How To Fly A Balloon; Balloon Digest.

Objective. To determine that the applicant exhibits instructional knowledge of the elements of the principles of flight by describing:

1. Physical laws applicable to balloon flight.
2. Effects of changes in temperature and density altitude on maintaining equilibrium.
3. Effects of false lift during takeoff, landing, and windshear penetration.

D. TASK: REGULATIONS AND PUBLICATIONS

REFERENCES: 14 CFR parts 1, 61, 91; NTSB 830; AC 00-2; AIM, Balloon Flight Manual.

Objective. To determine that the applicant exhibits instructional knowledge of the elements of regulations and publications, their purpose, general content, availability, and how to obtain revisions by describing:

1. 14 CFR parts 1, 61, 91 and NTSB 830.
2. Flight information publications.
3. Advisory circulars.
4. Practical test standards.
5. Balloon flight manual.

E. TASK: NATIONAL AIRSPACE SYSTEM

REFERENCES: 14 CFR part 91; AC 91-71; AIM.

Objective. To determine that the applicant exhibits instructional knowledge of the elements of the national airspace system by describing:

1. Definitions and dimensions of Class A, B, C, D, E, and G airspace.
2. Pilot certification, weather, and equipment requirements for operating in Class A, B, C, D, E, and G airspace.
3. Special use airspace and other airspace areas.

F. TASK: LOGBOOK ENTRIES AND CERTIFICATE ENDORSEMENTS

REFERENCES: 14 CFR part 61; AC 61-65, AC 61-98, AC 61-101.

Objective. To determine that the applicant exhibits instructional knowledge of the elements of logbook entries and certificate endorsements by describing:

1. Required logbook entries for instruction given.
2. Logbook entry certifying student's completion of presolo knowledge test.
3. Required student pilot certificate endorsements and appropriate logbook entries.
4. Preparation of a recommendation for a pilot practical test, including appropriate logbook entry.
5. Required endorsement of a pilot logbook for satisfactory completion of the required FAA flight review.
6. Instructor record keeping.

III. AREA OF OPERATION: PREFLIGHT PREPARATION

A. TASK: CERTIFICATES AND DOCUMENTS

REFERENCES: 14 CFR parts 43, 61, 91; AC 61-23; Balloon Flight Manual.

Objective. To determine that the applicant

1. Exhibits commercial pilot knowledge of the elements of certificates and documents by explaining:
2. Requirements for the issuance of pilot certificates and ratings, and the privileges and limitations of those certificates and ratings.
3. Medical requirements.
4. Airworthiness and registration certificates.
5. Balloon flight manuals.
6. Balloon maintenance/inspection requirements and associated records.

B. TASK: WEATHER INFORMATION

REFERENCES: AC 00-6, AC 00-45, AC 61-23, AC 61-84; AIM.

Objective. To determine that the applicant exhibits commercial pilot knowledge of the elements of weather information by explaining:

1. Importance of a thorough weather check.
2. Sources available for obtaining weather information.
3. Use of weather reports, forecasts, and charts.
4. Use of PIREP's, SIGMET's, and AIRMET's.
5. Recognition of aviation weather hazards and their effects on balloon operations.
6. Factors to be considered in making a "go/no go" decision.

C. TASK: FLIGHT PLANNING

REFERENCES: AC 61-21, AC 61-23, AC 61-84; Navigation Charts; AIM; NOTAM's.

Objective. To determine that the applicant:

1. Exhibits commercial pilot knowledge of the elements of flight planning by presenting and explaining a preplanned flight of maximum duration, appropriate to the ballooXBn used for the flight test, as previously assigned by the examiner. The final flight plan shall include real-time weather.

2. Uses appropriate, current aeronautical charts and appropriate, current local road/street maps.
3. Constructs a flight profile and plots a course for intended route of flight based on winds aloft.
4. Selects appropriate VHF communication frequencies, if radio equipped.
5. Identifies airspace, obstructions, and terrain features.
6. Selects suitable landing areas.
7. Extracts and applies pertinent information from AIM and NOTAM's, as appropriate.

D. TASK: PERFORMANCE AND LIMITATIONS

REFERENCES: AC 61-84; Balloon Flight Manual.

Objective. To determine that the applicant exhibits commercial pilot knowledge of the elements of performance and limitations, appropriate to the balloon used for the practical test, by explaining:

1. Use of performance charts, tables, and other data in determining performance in various phases of flight.
2. Computation of operating weight, maximum load, fuel quantity and endurance.
3. Determination of normal and maximum rates of ascent and descent, and the altitude required to arrest a high rate of descent.
4. Determination of envelope temperatures, including never-exceed temperature and maximum continuous temperature.
5. Effects of atmospheric conditions on performance.
6. Factors to be considered in determining that the required performance is within the balloon's capabilities and limitations.

E. TASK: OPERATION OF SYSTEMS

REFERENCE: Balloon Flight Manual.

Objective. To determine that the applicant exhibits commercial pilot knowledge of the elements of the operation of systems, as applicable to the balloon used for the practical test, by explaining:

1. Fuel system, burners, pilot lights, and associated gauges.
2. Flight instruments and gauges.
3. Venting and/or deflation systems.
4. Avionics/communications system, as appropriate.

IV. AREA OF OPERATION: PREFLIGHT LESSON ON A MANEUVER TO BE PERFORMED IN FLIGHT

NOTE: Examiner will select at least one maneuver from AREAS OF OPERATION VII through XI and ask the applicant to present a preflight lesson on the maneuver selected as the lesson would be taught to a student. Previously developed lesson plans from the applicant's library may be used.

TASK: MANEUVER LESSON

REFERENCES: AC 60-14; Balloon Digest; How To Fly A Balloon; Balloon Flight Manual.

Objective. To determine that the applicant exhibits instructional knowledge of the selected maneuver by:

1. Using a lesson plan that includes all essential items to make an effective and organized presentation.
2. Stating the objective.
3. Giving an accurate, comprehensive oral description of the maneuver, including the elements and associated common errors.
4. Using instructional aids, as appropriate.
5. Describing the recognition, analysis, and correction of common errors.

V. AREA OF OPERATION: PREFLIGHT PROCEDURES

A. TASK: LAUNCH SITE SELECTION

REFERENCES: AC 91-71; How To Fly A Balloon.

Objective. To determine that the applicant:

1. Exhibits commercial pilot knowledge by explaining the elements of launch site selection.
2. Arranges to launch at a suitable time, considering atmospheric conditions.
3. Selects a launch site with emphasis on—
    a. size and surface condition of site.
    b. consideration of accessibility and obstacles.
    c. surface wind and winds aloft.
    d. obstacles in vicinity of launch site.
    e. consideration of suitable landing areas based on wind conditions.
4. Makes a competent “go/no-go” decision considering all of the factors involved in the selection of a launch site.

B. TASK: CREW BRIEFING AND PREPARATION

REFERENCE: How To Fly A Balloon.

Objective. To determine that the applicant:

1. Exhibits commercial pilot knowledge by explaining the elements of crew briefing and preparation.
2. Designates a crew chief, if appropriate, and explains duties and responsibilities to each crewmember.
3. Briefs crewmembers in all areas of the flight, including layout and assembly; tie-off, if appropriate; inflation; in-flight; landing; recovery; and emergency procedures.
4. Establishes a common means of communication such as hand signals and/or two-way radio.
5. Describes the proposed direction of flight and estimated time aloft.
6. Ensures that all necessary equipment is on board.
7. Supervises and coordinates all activities.

C. TASK: LAYOUT AND ASSEMBLY

REFERENCE: Balloon Flight Manual.

Objective.
To determine that the applicant:

1. Exhibits commercial pilot knowledge by explaining the elements of layout and assembly.
2. Positions envelope and basket properly, considering wind conditions, surface, and obstacles.
3. Assembles fuel system (as appropriate) and checks for security, leaks, and correct fuel pressure.
4. Completes attachment of all cables and lines and assembles basket to envelope in accordance with the flight manual.
5. Completes an appropriate checklist.

D. TASK: PREFLIGHT INSPECTION

REFERENCES: How To Fly A Balloon; Balloon Flight Manual.

Objective. To determine that the applicant:

1. Exhibits commercial pilot knowledge by explaining the elements of a preflight inspection. This shall include the items which must be inspected, reasons for inspecting each item, and how to detect possible defects.
2. Inspects the balloon with reference to the checklist emphasizing the—
    a. basket and envelope, to include suspension and handling lines.
    b. venting and/or deflation systems.
    c. burner and fuel system check.
    d. instruments and gauges.
3. Verifies the balloon is in condition for safe flight.
4. Completes an appropriate checklist.

E. TASK: INFLATION

REFERENCES: How To Fly A Balloon; Balloon Flight Manual.

Objective. To determine that the applicant:

1. Exhibits commercial pilot knowledge by explaining the elements of inflation.
2. Accomplishes proper tie-off procedure, if appropriate.
3. Positions inflator for initial cold inflation.
4. Begins ignition and hot air inflation.

5. Inflates balloon to a vertical position.
6. Positions and secures the vent/deflation lines.
7. Completes an appropriate checklist.

F. TASK: BASKET/GONDOLA MANAGEMENT

REFERENCE: Balloon Flight Manual.

Objective. To determine that the applicant:

1. Exhibits commercial pilot knowledge by explaining the elements of basket/gondola management.
2. Ensures all loose items in the basket/gondola are secured.
3. Briefs passengers on the proper boarding, in-flight, and landing behavior and procedures.
4. Organizes material and equipment in a logical, efficient manner.

G. TASK: PRE-LAUNCH CHECK

REFERENCE: Balloon Flight Manual.

Objective. To determine that the applicant:

1. Exhibits commercial pilot knowledge by explaining the elements of a pre-launch check.
2. Reviews wind conditions, temperatures, and obstructions.
3. Performs final instrument check.
4. Ensures that vent/deflation lines are positioned and secured properly.
5. Accomplishes final coordination with ground crew, including signals and emergency procedures.
6. Accomplishes pre-launch checklist and confirms that the balloon is in safe operating condition.
7. Brings balloon to equilibrium.
8. Divides attention in and around the basket/gondola. Ensures no conflict with traffic prior to launch.

VI. AREA OF OPERATION: AIRPORT OPERATIONS

TASK: RADIO COMMUNICATIONS

REFERENCES: AC 61-21, AC 61-23; AIM.

Objective. To determine that the applicant exhibits commercial pilot knowledge of the elements of radio communications by explaining:

1. Selection and use of appropriate radio frequencies.
2. Recommended procedure and phraseology for radio voice communications.
3. Receipt, acknowledgment of, and compliance with, ATC clearances and other instructions.

VII. AREA OF OPERATION: LAUNCHES AND LANDINGS

A. TASK: NORMAL LAUNCH

REFERENCES: AC 91-71; How To Fly A Balloon; Balloon Flight Manual.

Objective. To determine that the applicant:

1. Exhibits commercial pilot knowledge by explaining the elements of a normal launch.
2. Directs ground crew to clear the area.
3. Recognizes equilibrium.
4. Uses tie-off-quick release line correctly, if appropriate.
5. Recognizes wind conditions and presence of false lift.
6. Appropriately controls lift-off and initial ascent.

B. TASK: LAUNCH OVER OBSTACLE

REFERENCES: AC 91-71; Balloon Digest.

NOTE: If conditions do not allow an additional launch to be performed, the applicant's knowledge of this TASK shall be evaluated orally.

Objective. To determine that the applicant:

1. Exhibits commercial pilot knowledge by explaining the elements of a launch over an obstacle.
2. Determines height of the obstacle.
3. Considers distance to the obstacle relative to the wind conditions.
4. Recognizes presence of false lift.
5. Lifts off and acts decisively so as to clear the obstacle safely.

C. TASK: APPROACH TO LANDING

REFERENCES: AC 91-71; How To Fly A Balloon; Balloon Flight Manual.

Objective. To determine that the applicant:

1. Exhibits commercial pilot knowledge by explaining the elements of an approach to landing.
2. Considers the wind conditions, landing area, obstructions, and surface, and selects the most suitable touchdown point.
3. Stows loose articles and secures equipment, as appropriate.

4. Ensures that each occupant is thoroughly briefed and positioned properly in accordance with landing conditions.
5. Check fuel tanks for quantity and selection and completes an appropriate checklist.
6. Establishes the appropriate approach profile and rate(s) of descent.
7. Makes a timely decision to abort the approach, if necessary.

D. TASK: STEEP APPROACH TO LANDING

REFERENCES: AC 91-71; How To Fly A Balloon; Balloon Flight Manual.

Objective. To determine that the applicant:

1. Exhibits commercial pilot knowledge by explaining the elements of a steep approach to landing.
2. Selects a landing site relative to wind speed and direction.
3. Briefs occupants and secures equipment.
4. Uses vents and burner controls properly to land balloon and control ground travel.
5. Exhibits timing, judgment, and control throughout the approach and landing.
6. Aborts landing, if requested by examiner.

E. TASK: NORMAL LANDING

REFERENCES: How To Fly A Balloon; Balloon Flight Manual.

Objective. To determine that the applicant:

1. Exhibits commercial pilot knowledge by explaining the elements of a normal landing.
2. Prepares vent/deflation system for use.
3. Touches down within the selected area or aborts the landing and ascends, if requested by examiner.
4. Uses burner controls and vent/deflation system properly to stabilize balloon on landing.
5. Stabilizes balloon prior to occupants exiting.

F. TASK: HIGH-WIND LANDING

REFERENCES: How To Fly A Balloon; Balloon Flight Manual.

NOTE: If a high-wind condition does not exist, the applicant's knowledge of this TASK shall be evaluated orally.

Objective. To determine that the applicant:

1. Exhibits commercial pilot knowledge by explaining the elements of a high-wind landing.
2. Ensures a thorough briefing to include positioning of occupants and securing of equipment.
3. Identifies hazards associated with a high-wind landing.
4. Selects a landing site appropriate for high-wind conditions.
5. Prepares vent/deflation system for use.
6. Uses burner controls and vent/deflation system to land the balloon and control ground travel.
7. Touches down within the selected area or aborts the landing and ascends, if requested by examiner.
8. Extinguishes pilot lights and shuts off fuel, as appropriate.

VIII. AREA OF OPERATION: PERFORMANCE MANEUVERS

A. TASK: ASCENTS

REFERENCES: AC 91-71; How To Fly A Balloon

Objective. To determine that the applicant:

1. Exhibits commercial pilot knowledge by explaining the elements of ascents.
2. Transitions from level flight to ascent, as specified by the examiner.
3. Ascends at a specified rate, ±50 feet (20 meters) per minute.
4. Transitions from ascent to level flight at an altitude specified by the examiner, ±50 feet (20 meters)

B. TASK: ALTITUDE CONTROL

REFERENCES: AC 91-71; How To Fly A Balloon.

Objective. To determine that the applicant:

1. Exhibits commercial pilot knowledge by explaining the elements of altitude control.
2. Recognizes vertical movement.
3. Maintains equilibrium by smooth use of burner controls.
4. Uses instruments to assist in altitude control.
5. Maintains assigned altitudes, ±50 feet (20 meters.

C. TASK: DESCENTS

REFERENCES: AC 91-71; How To Fly A Balloon.

Objective. To determine that the applicant:

1. Exhibits commercial pilot knowledge by explaining the elements of descents.
2. Transitions from level flight to descent, as specified by the examiner.
3. Descends at a specified rate, ±50 feet (20 meters) per minute.
4. Transitions from descent to level flight at an altitude specified by the examiner, ±50 feet (20 meters).

D. TASK: RAPID ASCENT AND DESCENT

REFERENCE: Balloon Flight Manual.

Objective. To determine that the applicantexhibits commercial pilot knowledge of the elements of a rapid ascent and descent by explaining:

1. Situations requiring use of a rapid ascent and descent.
2. Exceeding manufacturer's limitations.
3. Potential problems with envelope distortions.
4. Time and altitude required to recover from a rapid descent.
5. Reasons for monitoring temperature control during a rapid ascent and descent.

E. TASK: CONTOUR FLYING (LBH)

REFERENCE: How To Fly A Balloon.

Objective. To determine that the applicant:

1. Exhibits commercial pilot knowledge by explaining the elements of contour flying.
2. Uses controls properly to maintain the altitude desired based on appropriate clearance over terrain, obstacles and consistent with safety.
3. Considers effects of wind gusts, wind shear, thermal activity and orographic conditions.
4. Avoids overburning and overventing.
5. Divides attention between balloon control, ground track, and visual scanning.

F. TASK: HIGH ALTITUDE FLIGHT (LBG)

REFERENCE: Balloon Flight Manual.

Objective. To determine that the applicant exhibits commercial pilot knowledge of the elements of high altitude flight by explaining:

1. Regulatory requirements for use of oxygen.
2. Physiological effects of high altitude flight.
3. Effects of high altitude on fuel system and performance.
4. Density altitude and its effects on flight characteristics.
5. Difficulties associated with altitude control.

G. TASK: OBSTACLE AVOIDANCE (LBH)

REFERENCE: Balloon Flight Manual.

Objective. To determine that the applicant exhibits commercial pilot knowledge of the elements of obstacle avoidance by explaining:

1. Importance of timely recognition of obstacles, particularly powerlines.
2. Techniques that can be used to avoid these obstacles.
3. Proper procedures to be used if collision is imminent.

H. TASK: TETHERING (LBH)

REFERENCE: Balloon Flight Manual.

Objective. To determine that the applicant exhibits commercial pilot knowledge of the elements of tethering by explaining:

1. Recommended procedures to include number, strength, and location of lines in accordance with flight manual.
2. Size of area required, considering wind conditions and obstructions.
3. Effects of false lift and wind gusts.
4. Importance of briefing ground crew on procedures, to include crowd control.

I. TASK: WINTER FLYING

REFERENCES: How To Fly A Balloon; Balloon Flight Manual.

Objective. To determine that the applicant exhibits commercial pilot knowledge of the elements of winter flying by explaining:

1. Fuel pressure concerns and proper methods of pressurizing fuel tanks.
2. Equipment and preparation necessary for cold temperature operations.
3. Added concerns for fuel vaporization, leaks, and/or fire risk.

J. TASK: MOUNTAIN FLYING

REFERENCES: How To Fly A Balloon; Balloon Flight Manual.

Objective. To determine that the applicant exhibits commercial pilot knowledge of the elements of mountain flying by explaining:

1. Proper preparation, equipment, and survival supplies necessary for flight over mountainous terrain.
2. Availability of and accessibility to landing areas.
3. Recognition of cloud formations and descending air currents on the leeward side of mountains as evidence of possible turbulence.
4. Caution required regarding windshear encounters and possible rapid weather changes.

IX. AREA OF OPERATION: NAVIGATION

TASK: NAVIGATION

REFERENCES: AC 61-21, AC 61-23.

Objective. To determine that the applicant:

1. Exhibits commercial pilot knowledge by explaining the elements of navigation.
2. Identifies airspace and altitude restrictions (if appropriate).
3. Identifies landmarks by relating surface features to chart symbols.
4. Verifies balloon's position at all times.
5. Manages fuel properly.
6. Determines duration of the flight, considering—
   a. availability of suitable landing areas.
   b. fuel consumption.
   c. wind and other atmospheric conditions.
   d. obstacles.
7. Notes differences between planned flight and the actual flight.

X. AREA OF OPERATION: EMERGENCY OPERATIONS

A. TASK: SYSTEMS AND EQUIPMENT MALFUNCTIONS

REFERENCE: Balloon Flight Manual.

Objective. To determine that the applicant exhibits commercial pilot knowledge of the elements of systems and equipment malfunctions, appropriate to the balloon used for the practical test, by explaining recommended pilot action for:

1. Pilot light flameout or failure.
2. Blast valve failure.
3. Fuel exhaustion.
4. Propane leak and/or fire.
5. Any other malfunction that may occur.

B. TASK: EMERGENCY EQUIPMENT AND SURVIVAL GEAR

REFERENCES: AC 61-21; Balloon Flight Manual.

Objective. To determine that the applicant exhibits commercial pilot knowledge of the elements of emergency equipment and survival gear, appropriate to the balloon used for the practical test, by explaining:

1. Location and purpose.
2. Method of operation or use.
3. Equipment appropriate for operation in various climates and types of terrain.

C. TASK: WATER LANDING

REFERENCE: Balloon Flight Manual.

Objective. To determine that the applicant exhibits commercial pilot knowledge of the elements of a water landing by explaining:

1. The emergency conditions under which water landings are necessary.
2. Consideration for wind effects and water current.
3. Preparation required for contact with water, to include briefing passengers.
4. Procedure to be used after water contact.

D. TASK: THERMAL FLIGHT

REFERENCE: Balloon Flight Manual.

Objective. To determine that the applicant exhibits commercial pilot knowledge of the elements of thermal flight by explaining:

1. Conditions that can cause thermal activity.
2. Recognition and effects of thermal activity on balloon flight.
3. Procedures that can be followed when encountering thermal activity.

XI. AREA OF OPERATION: POSTFLIGHT PROCEDURES

A. TASK: RECOVERY

REFERENCES: How To Fly A Balloon; Balloon Flight Manual.

Objective. To determine that the applicant:

1. Exhibits commercial pilot knowledge by explaining the elements of a recovery.
2. Coordinates recovery with landowner, as appropriate.
3. Supervises ground crew during recovery, including vehicle and spectator control.
4. Ensures importance of minimizing property damage.

B. TASK: DEFLATION AND PACK-UP

REFERENCES: Balloon Digest; Balloon Flight Manual.

Objective. To determine that the applicant:

1. Exhibits commercial pilot knowledge by explaining the elements of deflation and pack-up.
2. Ensures the fuel system is secure.
3. Deflates envelope properly, considering wind conditions and obstacles.
4. Disassembles envelope and basket components, as appropriate.
5. Packs and stores envelope, basket and components, and fuel system, as appropriate.
6. Performs satisfactory postflight inspection.
7. Completes an appropriate checklist.

C. TASK: REFUELING (LBH)

REFERENCES: Balloon Digest; Balloon Flight Manual.

Objective. To determine that the applicant exhibits commercial pilot knowledge of the elements of refueling by explaining:

1. Physical properties of propane.
2. Propane cylinders and related parts.
3. Safety factors, to include ventilation.
4. Danger of explosion and burns.
5. Moisture contamination.
6. Proper method of filling cylinders.

SECTION 2

COMMERCIAL PILOT LIGHTER-THAN-AIR — AIRSHIP

Practical Test Standards

CONTENTS

Lighter-Than-Air — Airship

Addition of a Lighter-Than-Air/Airship rating to an existing Commercial Pilot Certificate

Required TASKS are indicated by either the TASK letter(s) that apply(s) or an indication that all or none of the TASKS must be tested.						
Area of Operation	Rating(s) Held					
	ASE	AME	RH	RG	G	LB
I	ALL	ALL	ALL	ALL	ALL	NONE
II	B,C,D,E, F	B,C,D,E F	B,C,D,E F	B,C,D,E F	B,C,D,E F	B,C,D,E, F
III	D,E	D,E	D,E	D,E	D,E	B,C,D,E
IV	ALL	ALL	ALL	ALL	ALL	NONE
V	ALL	ALL	ALL	ALL	ALL	ALL
VI	B	B	B	B	ALL	ALL
VII	ALL	ALL	ALL	ALL	ALL	ALL
VIII	ALL	ALL	ALL	ALL	ALL	ALL
IX	NONE	NONE	NONE	NONE	ALL	ALL
X	ALL	ALL	ALL	ALL	ALL	ALL
XI	ALL	ALL	ALL	ALL	ALL	ALL

LEGEND

ASE	Airplane Single-Engine
AME	Airplane Multiengine
RH	Rotorcraft Helicopter
RG	Rotorcraft Gyroplane
G	Glider
LB	Lighter-Than-Air Balloon

If an applicant holds more than one rating on a commercial pilot certificate and the table indicates both ALL and NONE for a particular AREA OF OPERATION, the NONE entry applies. This is logical since the applicant has satisfactorily accomplished the AREA OF OPERATION on a previous commercial pilot practical test. If an applicant holds a flight Instructor certificate, AREAS OF OPERATION I and IV are not required. However, at the discretion of the examiner, the applicant's competence in all AREAS OF OPERATION may be evaluated.

APPLICANT'S PRACTICAL TEST CHECKLIST (AIRSHIP)

APPOINTMENT WITH EXAMINER

EXAMINER'S NAME____

LOCATION ____

DATE/TIME____

ACCEPTABLE AIRCRAFT

- ○ Aircraft Documents:
  - Airworthiness Certificate
  - Registration Certificate
  - Operating Limitations
- ○ Aircraft Maintenance Records:
  - Airworthiness Inspections
- ○ Pilot's Operating Handbook, FAA-Approved Flight Manual

PERSONAL EQUIPMENT

- ○ View-Limiting Device
- ○ Practical Test Standard
- ○ Current Aeronautical Charts
- ○ Computer and Plotter
- ○ Flight Plan and Flight Log Forms
- ○ Current AIM and Airport Facility Directory

PERSONAL RECORDS

- ○ Identification - Photo/Signature ID
- ○ Pilot Certficate
- ○ Current Medical Certificate
- ○ Completed FAA Form 8710-1, Airman Certificate and/or Rating Application

- AC Form 8080-2, Airman Written Test Report or Computer Test Report
- Pilot Logbook with Appropriate Instructor Endorsements
- FAA Form 8060-5, Notice of Disapproval (if applicable)
- Approved School Graduation Certificate (if applicable)
- Examiner's Fee (if applicable)

EXAMINER'S PRACTICAL TEST CHECKLIST (AIRSHIP)

APPLICANT'S NAME ______________________________

LOCATION ______________________________

DATE/TIME ______________________________

I. FUNDAMENTALS OF INSTRUCTING

- ○ A. The Learning Process
- ○ B. Human Behavior
- ○ C. The Teaching Process
- ○ D. Teaching Methods
- ○ E. Critique and Evaluation
- ○ F. Flight Instructor Characteristics and Responsibilities
- ○ G. Planning Instructional Activity

II. TECHNICAL SUBJECTS

- ○ A. Aeromedical Factors
- ○ B. Visual Scanning and Collision Avoidance
- ○ C. Use of Distractions During Flight Training
- ○ D. Principles of Flight
- ○ E. Airship Weigh-off, Ballast, and Trim
- ○ F. Night Operations
- ○ G. Regulations and Publications
- ○ H. National Airspace System
- ○ I. Logbook Entries and Certificate Endorsements

III. PREFLIGHT PREPARATION

- ○ A. Certificates and Documents
- ○ B. Weather Information

- C. Cross-Country Flight Planning
- D. Performance and Limitations
- E. Operation of Systems

IV. PREFLIGHT LESSON ON A MANEUVER TO BE PERFORMED IN FLIGHT

- Maneuver Lesson

V. PREFLIGHT PROCEDURES

- A. Preflight Inspection
- B. Cockpit Management
- C. Engine Starting
- D. Unmasting and Positioning for Takeoff
- E. Ground Handling
- F. Before Takeoff Check

VI. AIRPORT OPERATIONS

- A. Radio Communications
- B. Traffic Pattern Operations
- C. Airport, Runway, and Taxiway Markings and Lighting

VII. TAKEOFFS, LANDINGS, AND GO-AROUND

- A. Ground Weigh-Off
- B. Up-Ship Takeoff
- C. Wheel Takeoff
- D. Approach and Landing
- E. Go-Around

VIII. PERFORMANCE MANEUVERS

- A. Flight To, From, and At Pressure Height
- B. In-Flight Weigh-Off

- C. Manual Pressure Control
- D. Static and Dynamic Trim

IX. NAVIGATION

- A. Pilotage and Dead Reckoning
- B. Diversion
- C. Lost Procedures
- D. Navigation Systems and ATC Radar Services

X. EMERGENCY OPERATIONS

- A. Aborted Takeoff
- B. Engine Failure During Takeoff
- C. Engine Failure During Flight
- D. Engine Fire During Flight
- E. Envelope Emergencies
- F. Free Ballooning
- G. Ditching and Emergency Landing
- H. Systems and Equipment Malfunctions

XI. POSTFLIGHT PROCEDURES

- A. Masting
- B. Post-masting

I. AREA OF OPERATION: FUNDAMENTALS OF INSTRUCTING

NOTE: The examiner will select TASK F and at least one other TASK.

A. TASK: THE LEARNING PROCESS

REFERENCE: AC 60-14.

Objective. To determine that the applicant exhibits instructional knowledge of the elements of the learning process by describing:

1. The definition and characteristics of learning.
2. Practical application of the laws of learning.
3. Factors involved in how people learn.
4. Recognition and proper use of the various levels of learning.
5. Principles that are applied in learning a skill.
6. Factors of forgetting and retention.
7. How the transfer of learning affects the learning process.
8. How the formation of habit patterns affects the learning process.

B. TASK: HUMAN BEHAVIOR

REFERENCE: AC 60-14.

Objective. To determine that the applicant exhibits instructional knowledge of the elements of human behavior by describing:

1. Control of human behavior.
2. Development of student potential.
3. Relationship of human needs to behavior and learning.
4. Relationship of defense mechanisms to student learning and pilot decision making.
5. General rules which a flight instructor should follow during student training to ensure good human relations.

C. TASK: THE TEACHING PROCESS

REFERENCE: AC 60-14.

Objective. To determine that the applicant exhibits instructional knowledge of the elements of the teaching process by describing:

1. Preparation of a lesson for a ground or flight instructional period.
2. Presentation of knowledge and skills, including the methods which are suitable in particular situations.
3. Application, by the student, of the knowledge and skills presented by the instructor.
4. Review of the material presented and the evaluation of student performance and accomplishment.

D. TASK: TEACHING METHODS

REFERENCE: AC 60-14.

Objective. To determine that the applicant exhibits instructional knowledge of the elements of teaching methods by describing:

1. The organization of a lesson, i.e., introduction, development, and conclusion.
2. The lecture method.
3. The guided discussion method.
4. The demonstration-performance method.
5. Computer/video assisted instruction.

E. TASK: CRITIQUE AND EVALUATION

REFERENCE: AC 60-14.

Objective. To determine that the applicant exhibits instructional knowledge of the elements of critique and evaluation by describing:

1. Purpose and characteristics of an effective critique.
2. Difference between critique and evaluation.
3. Characteristics of effective oral questions and what type to avoid.
4. Responses to student questions.
5. Characteristics and development of effective written tests.
6. Characteristics and uses of performance tests, specifically, the FAA practical test standards.

F. TASK: FLIGHT INSTRUCTOR CHARACTERISTICS AND RESPONSIBILITIES

REFERENCE: AC 60-14.

Objective. To determine that the applicant exhibits instructional knowledge of the elements of flight instructor characteristics and responsibilities by describing:

1. Characteristics and qualifications of a professional flight instructor.
2. Role of the flight instructor in dealing with student stress, anxiety, and psychological abnormalities.
3. Flight instructor's responsibility with regard to student pilot supervision and surveillance.
4. Flight instructor's authority and responsibility for endorsements and recommendations.
5. Flight instructor's responsibility in the conduct of the required FAA flight review.

G. TASK: PLANNING INSTRUCTIONAL ACTIVITY

REFERENCE: AC 60-14.

Objective. To determine that the applicant exhibits instructional knowledge of the elements of planning instructional activity by describing:

1. Development of a course of training.
2. Content and use of a training syllabus.
3. Purpose, characteristics, proper use, and items of a lesson plan.
4. Flexibility features of a course of training, syllabus, and lesson plan required to accommodate students with varying backgrounds, levels of experience, and ability.

II. AREA OF OPERATION: TECHNICAL SUBJECTS

A. TASK: AEROMEDICAL FACTORS

REFERENCES: AC 61-21; AIM.

Objective. To determine that the applicant exhibits instructional knowledge of the elements of aeromedical factors by describing:

1. How to obtain an appropriate medical certificate.[1]
2. Hypoxia, its symptoms, effects, and corrective action.
3. Hyperventilation, its symptoms, effects, and corrective action.
4. Middle ear and sinus problems, their causes, effects, and corrective action.
5. Spatial disorientation, its causes, effects, and corrective action.
6. Motion sickness, its causes, effects, and corrective action.
7. Effects of alcohol and drugs, and their relationship to safety.
8. Carbon monoxide poisoning, its symptoms, effects, and corrective action.
9. Effect of nitrogen excess during scuba dives and how this affects a pilot and passengers during flight.
10. Fatigue, its effects and corrective action.

B. TASK: VISUAL SCANNING AND COLLISION AVOIDANCE

REFERENCES: AC 61-21, AC 90-48; AIM.

Objective. To determine that the applicant exhibits instructional knowledge of the elements of visual scanning and collision avoidance by describing:

1. Relationship between a pilot's physical or mental condition and vision.
2. Practice of "time sharing" of attention inside and outside the cockpit.
3. Proper visual scanning technique.
4. Various optical illusions.
5. Relationship between poor visual scanning habits and increased collision risk.
6. Appropriate clearing procedures.
7. Situations which involve the greatest collision risk.

---

1 The flight instructor should encourage a person considering flight training to obtain an appropriate medical certificate from an Aviation Medical Examiner before training is started.

C. TASK: USE OF DISTRACTIONS DURING FLIGHT TRAINING

REFERENCE: AC 61-67.

Objective. To determine that the applicant exhibits instructional knowledge of the elements of the use of distractions by describing:

1. Flight situations where pilot distraction can be a cause factor of aircraft accidents.
2. Selection of realistic distractions for specific flight situations.
3. Relationship between division of attention and flight instructor use of distractions.
4. Difference between proper use of distractions and harassment.

D. TASK: PRINCIPLES OF FLIGHT

REFERENCE: Aerostatics.

Objective. To determine that the applicant exhibits instructional knowledge of the elements of the principles of flight by describing:

1. Aerostatics—
    a. physical properties of gases.
    b. laws of Charles, Boyle, and Archimedes.
    c. application of these laws (pressure height, superheat, buoyancy).
    d. lift (gross, net, useful, disposable).
2. Aerodynamics—
    a. fineness ratio.
    b. aerodynamic pressure.
    c. dynamic lift/drag.

E. TASK: AIRSHIP WEIGH-OFF, BALLAST, AND TRIM

REFERENCE: Airship Flight Manual.

Objective. To determine that the applicant exhibits instructional knowledge of the elements of airship weigh-off, ballast, and trim by describing:

1. Purpose of and procedure for weigh-off.
2. Effects of ballast and trim on center of buoyancy and performance.
3. Methods of ballasting and trim control.
4. Effects of temperature, pressure, and humidity prior to and during flight.
5. Determination of total weight and changes that occur due to fuel consumption.

F. TASK: NIGHT OPERATIONS

REFERENCES: AC 61-21, AC 61-23; AIM.

Objective. To determine that the applicant exhibits instructional knowledge of the elements of night operations by describing:

1. Factors of night vision, disorientation, and optical illusions.
2. Weather considerations specific to night operations.
3. Thorough preflight inspection to include lighting.
4. Proper adjustment of interior lights, including availability of flashlight.
5. Crew briefing, including hand/voice signals and crew lighting system.
6. Engine starting procedures, including proper use of exterior lighting prior to start.
7. Unmasting, takeoff, in-flight orientation, landing, and go-around.
8. Importance of verifying the airship's attitude by visual reference to flight instruments.
9. Emergencies such as electrical failure and engine malfunction.
10. Traffic patterns.
11. Masting, engine shutdown, and postflight procedures.

G. TASK: REGULATIONS AND PUBLICATIONS

REFERENCES: 14 CFR parts 1, 61, 91; NTSB 830; AC 00-2; AIM; Airship Flight Manual.

Objective: To determine that the applicant exhibits instructional knowledge of the elements of regulations and publications, their purpose, general content, availability, and how to obtain revisions by describing:

1. 14 CFR parts 1, 61, 91 and NTSB 830.
2. Flight information publications.
3. Advisory circulars.
4. Practical test standards.
5. Airship flight manual.

H. TASK: NATIONAL AIRSPACE SYSTEM

REFERENCES: 14 CFR part 91; AIM.

Objective. To determine that the applicant exhibits instructional knowledge of the elements of the national airspace system by describing:

1. Definition and dimensions of Class A, B, C, D, E, and G airspace.
2. Pilot certification, weather, and equipment requirements for operating in Class A, B, C, D, E, and G airspace.
3. Special use airspace and other airspace areas.

I. TASK: LOGBOOK ENTRIES AND CERTIFICATE ENDORSEMENTS

REFERENCES: 14 CFR part 61; AC 61-65, AC 61-98, AC 61-101.

Objective. To determine that the applicant exhibits instructional knowledge of the elements of logbook entries and certificate endorsements by describing:

1. Required logbook entries for instruction given.
2. Logbook entry certifying student's completion of presolo knowledge test.
3. Required student pilot certificate endorsements, including appropriate logbook entries.
4. Preparation of a recommendation for a pilot practical test, including appropriate logbook entry.
5. Required endorsement of a pilot logbook for satisfactory completion of an FAA flight review.
6. Flight instructor record keeping.

III. AREA OF OPERATION: PREFLIGHT PREPARATION

A. TASK: CERTIFICATES AND DOCUMENTS

REFERENCES: 14 CFR parts 43, 61, 91; AC 61-23; Airship Flight Manual.

Objective. To determine that the applicant exhibits commercial pilot knowledge of the elements of certificates and documents by explaining:

1. Requirements for the issuance of pilot certificates and ratings, and the privileges and limitations of those certificates and ratings.
2. Class and duration of medical certificates.
3. Airworthiness and registration certificates.
4. Airship flight manuals.
5. Airship maintenance/inspection requirements and associated records.

B. TASK: WEATHER INFORMATION

REFERENCES: AC 00-6, AC 00-45, AC 61-23; AIM.

Objective. To determine that the applicant exhibits commercial pilot knowledge of the elements of weather information by analyzing weather reports, charts, and forecasts from various sources with emphasis on:

1. Importance of a thorough weather check.
2. Sources available for obtaining weather information.
3. Use of weather reports, forecasts, and charts.
4. Use of PIREP's, SIGMET's, and AIRMET's.
5. Recognition of aviation weather hazards to include wind shear.
6. Factors to be considered in making a "go/no-go" decision.

C. TASK: CROSS-COUNTRY FLIGHT PLANNING

REFERENCES: AC 61-21, AC 61-23, AC 61-84; Navigation Charts; Airport/Facility Directory; AIM.

Objective. To determine that the applicant:

1. Exhibits commercial pilot knowledge of the elements of cross-country flight planning by presenting and explaining a preplanned VFR cross-country flight of maximum duration, appropriate to the airship used for the flight test, as previously assigned by the examiner. The final flight plan shall include real-time weather to the first fuel stop, with maximum allowable passenger and baggage loads.

2. Uses appropriate, current aeronautical charts.
3. Plots a course for the intended route of flight, considering terrain and pressure height.
4. Identifies airspace, obstructions, and alternate airports.
5. Selects easily identifiable en route checkpoints.
6. Selects the most favorable altitudes, considering weather conditions and equipment capabilities.
7. Computes headings, flight time, and fuel requirements.
8. Selects appropriate navigation systems/facilities and communication frequencies.
9. Considers availability of facilities and ground crew at destination.
10. Extracts and applies pertinent information from AIM, Airport/Facility Directory, NOTAM's, and other flight publications.
11. Completes a navigation log and simulates filing a VFR flight plan.

D. TASK: PERFORMANCE AND LIMITATIONS

REFERENCE: Airship Flight Manual.

Objective. To determine that the applicant exhibits commercial pilot knowledge of the elements of performance and limitations by explaining:

1. Determination of weight and trim condition.
2. Use of performance charts, tables, and other data in determining performance in various phases of flight.
3. Effects of the following conditions on airship performance—
    a. weights and lift (static and dynamic).
    b. relationship of ballonet fullness to pressure height.
    c. superheat on percent of fullness.
    d. average ballonet volume with respect to total envelope volume.
    e. loss of gross lift when above pressure height.
    f. leaks in ballonets and envelope.
    g. gas purity on lift.
    h. temperature inversion on descents.
    i. superheat on lift.
    j. air temperature changes.
    k. humidity, altitude, and temperature on lift.
    l. maximum rate climb and descent limitations.

E. TASK: OPERATION OF SYSTEMS

REFERENCE: Airship Flight Manual.

Objective. To determine that the applicant exhibits commercial pilot knowledge of the elements of the operation of systems, as applicable to the airship used for the practical test, by explaining:

1. Surface control systems.
2. Flight instruments and associated controls.
3. Landing gear.
4. Engines and propellers.
5. Fuel and oil system.
6. Electrical system.
7. Envelope/ballonet pressure systems.
8. Avionics and auxiliary equipment.
9. Any system unique to the airship flown.

IV. AREA OF OPERATION: PREFLIGHT LESSON ON A MANEUVER TO BE PERFORMED IN FLIGHT

NOTE: Examiner will select at least one maneuver from AREAS OF OPERATION VII through XI, and ask the applicant to present a preflight lesson on the maneuver selected as the lesson would be taught to a student. Previously developed lesson plans from the applicant's library may be used.

TASK: MANEUVER LESSON

REFERENCES: AC 60-14; Airship Flight Manual.

Objective. To determine that the applicant exhibits instructional knowledge of the selected maneuver by:

1. Using a lesson plan that includes all essential items to make an effective and organized presentation.
2. Stating the objective.
3. Giving an accurate, comprehensive oral description of the maneuver, including the elements and associated common errors.
4. Using instructional aids, as appropriate.
5. Describing the recognition, analysis, and correction of common errors.

V. AREA OF OPERATION: PREFLIGHT PROCEDURES

A. TASK: PREFLIGHT INSPECTION

REFERENCES: AC 61-21; Airship Flight Manual.

Objective. To determine that the applicant:

1. Exhibits commercial pilot knowledge of the elements of a preflight inspection, as applicable to the airship used for the practical test, by explaining reasons for the visual inspection, what items should be inspected, and how defects are detected.
2. Inspects airship with reference to an appropriate checklist.
3. Verifies the airship is in condition for safe flight.

B. TASK: COCKPIT MANAGEMENT

REFERENCES: AC 61-21; Airship Flight Manual.

Objective. To determine that the applicant:

1. Exhibits commercial pilot knowledge by explaining the elements of cockpit management.
2. Briefs passengers on the use of safety belts and emergency procedures.
3. Organizes essential material and equipment in a manner that makes the items readily available.
4. Maintains orderly records reflecting progress of the flight, as appropriate.

C. TASK: ENGINE STARTING

REFERENCES: AC 61-21, AC 61-23, AC 91-13; Airship Flight Manual.

Objective. To determine that the applicant:

1. Exhibits commercial pilot knowledge by explaining the elements of engine starting, as applicable to the airship used for the practical test.
2. Observes safety precautions of starting, considering open hangars, other aircraft, and safety of nearby persons and property.
3. Considers starting under various atmospheric conditions and use of an external power source, if appropriate.
4. Accomplishes correct starting procedure inCoordinates with ground crew to minimize movement of airship during and after start.
5. Completes an appropriate checklist.

D. TASK: UNMASTING AND POSITIONING FOR TAKEOFF

REFERENCE: Airship Flight Manual.

Objective. To determine that the applicant:

1. Exhibits commercial pilot knowledge by explaining the elements of unmasting and positioning for takeoff.
2. Briefs ground crew and coordinates hand signals and voice commands.
3. Prevents airship from riding up on the mast.
4. Ensures proper envelope pressure and trim before coming off the mast.
5. Uses ground crew and airship controls properly to move away from the mast and into position for takeoff.
6. Divides attention inside and outside the cockpit so as to avoid possible immediate takeoff after coming off the mast.
7. Completes an appropriate checklist.

E. TASK: GROUND HANDLING

REFERENCE: Airship Flight Manual.

Objective. To determine that the applicant:

1. Exhibits commercial pilot knowledge by explaining the elements of ground handling, as applicable to the airship used for the practical test.
2. Determines required number of crew members, considering weather conditions, status of the airship, and method of handling.
3. Briefs ground crew on all pertinent phases of ground handling procedures.
4. Maintains coordination with crew chief and uses proper hand signals and voice commands with crew.
5. Recognizes undesirable airship movement and takes appropriate action.
6. Maintains proper envelope pressure and trim and alertness for wind shifts.
7. Maintains proper position while controlled by ground crew.

F. TASK: BEFORE TAKEOFF CHECK

REFERENCES: AC 61-21; Airship Flight Manual.

Objective. To determine that the applicant:

1. Exhibits commercial pilot knowledge by explaining the elements of a before takeoff check.
2. Positions airship properly to avoid hazards.
3. Divides attention inside and outside the cockpit.

4. Ensures that engine temperatures and pressures are suitable for runup and takeoff.
5. Accomplishes before takeoff checklist and confirms that airship is in safe operating condition.
6. Reviews takeoff performance, wind direction and speed, expected takeoff distance, emergency procedures, and the departure procedure.
7. Ensures that takeoff path is clear of obstacles.
8. Assures no conflict with traffic prior to takeoff.

VI. AREA OF OPERATION: AIRPORT OPERATIONS

A. TASK: RADIO COMMUNICATIONS

REFERENCES: AC 61-21, AC 61-23; AIM.

Objective. To determine that the applicant:

1. Exhibits commercial pilot knowledge by explaining the elements of radio communications.
2. Demonstrates radio communications by—
    a. selecting appropriate frequencies.
    b. transmitting using recommended phraseology.
    c. acknowledging and complying with ATC instructions.
3. Uses appropriate procedures for simulated radio communications failure.

B. TASK: TRAFFIC PATTERN OPERATIONS

REFERENCES: AC 61-21, AC 61-23; AIM.

Objective. To determine that the applicant:

1. Exhibits commercial pilot knowledge by explaining the elements of traffic patterns to include operations at controlled and uncontrolled airports, runway incursion and collision avoidance, wake turbulence avoidance, and wind shear.
2. Complies with traffic pattern instructions.
3. Maintains adequate spacing from other traffic.
4. Corrects for wind drift to maintain proper ground track.
5. Maintains orientation with runway or landing area to be used.
6. Establishes a final approach at an appropriate distance from runway or landing area.
7. Maintains appropriate traffic pattern altitude, ±100 feet (30 meters).
8. Maintains airspeed for current static condition of the airship.
9. Completes an appropriate checklist.

C. TASK: AIRPORT, RUNWAY, AND TAXIWAY MARKINGS AND LIGHTING

REFERENCES: AC 61-21, AC 61-23; AIM.

Objective. To determine that the applicant:

1. Exhibits commercial pilot knowledge by explaining the elements of airport, runway, and taxiway signs, markings, and lighting.
2. Identifies and interprets airport, runway, and taxiway signs, markings, and lighting.

VII. AREA OF OPERATION: TAKEOFFS, LANDINGS, AND GO-AROUNDS

A. TASK: GROUND WEIGH-OFF

REFERENCE: Airship Flight Manual.

Objective. To determine that the applicant:

1. Exhibits commercial pilot knowledge by explaining the elements of a ground weigh-off.
2. Determines static and trim condition.
3. Maintains zero inclination and heading into the wind.
4. Assists ground crew to minimize airship movement.
5. Checks weigh-off and trim with neutral elevators when HANDS OFF command is given.
6. Ballasts the airship according to the conditions and type of flight contemplated without exceeding weight limits.
7. Completes an appropriate checklist.

B. TASK: UP-SHIP TAKEOFF

REFERENCE: Airship Flight Manual.

Objective. To determine that the applicant:

1. Exhibits commercial pilot knowledge by explaining the elements of an up-ship takeoff.
2. Determines heaviness limitations and weather conditions under which an up-ship takeoff may be made.
3. Ensures that sufficient ground crew are available so as to obtain adequate upward velocity.
4. Idles engines and uses rudder as necessary during weigh-off.
5. Remains within takeoff heaviness limits.
6. Uses proper and timely hand signals and voice commands with ground crew.
7. Applies up-elevator pressure as ground crew lifts airship and transitions to a noseup attitude, keeping tail clear of the ground.
8. Applies power as airship nears the top of its upward thrust.
9. Prevents tail from striking the ground.
10. Increases airspeed sufficiently to carry the load dynamically.
11. Completes an appropriate checklist.

C. TASK: WHEEL TAKEOFF

REFERENCE: Airship Flight Manual.

Objective. To determine that the applicant:

1. Exhibits commercial pilot knowledge by explaining the elements of a wheel takeoff to include wheel takeoffs under various degrees of heaviness.
2. Determines approximate takeoff roll and ensures that area is clear and sufficient, considering wind conditions and field surface.
3. Positions airship to utilize the maximum available takeoff area and maintains trim.
4. Uses proper hand signals and voice commands with ground crew.
5. Applies power slowly, in a timely manner.
6. Attains sufficient airspeed to carry the load dynamically while on the wheel.
7. Uses elevators to assist airship in lifting dynamically.
8. Maintains directional control and proper inclination to keep tail off ground.
9. Completes an appropriate checklist.

D. TASK: APPROACH AND LANDING

REFERENCE: Airship Flight Manual.

Objective. To determine that the applicant:

1. Exhibits commercial pilot knowledge by explaining the elements of an approach and landing, to include light and heavy airships.
2. Accomplishes in-flight (static) weigh-off prior to commencing the approach.
3. Adjusts trim, as necessary, for landing, considering weight and condition of air.
4. Coordinates flight and power controls, as necessary.
5. Makes smooth and gradual approach maintaining direction and angle of descent.
6. Recognizes and adheres to waveoff signals.
7. Lands at a speed appropriate for approaching the ground crew.
8. Reverses thrust, if applicable.
9. Completes an appropriate checklist.

E. TASK: GO-AROUND

REFERENCE: Airship Flight Manual.

Objective. To determine that the applicant:

1. Exhibits commercial pilot knowledge by explaining the elements of a go-around.
2. Makes a timely decision to discontinue the approach to landing.
3. Uses correct procedures for a light or heavy airship, as appropriate.
4. Coordinates use of power and flight controls to effect a smooth transition to a climb attitude.
5. Completes an appropriate checklist.

VIII. AREA OF OPERATION: PERFORMANCE MANEUVERS

A. TASK: FLIGHT TO, FROM, AND AT PRESSURE HEIGHT

REFERENCE: Airship Flight Manual.

NOTE: If airship is unable to reach pressure height due to cloud conditions, this TASK may be evaluated through oral testing.

Objective. To determine that the applicant:

1. Exhibits commercial pilot knowledge by explaining the elements of flight to, at, and from pressure height.
2. Coordinates use of power and flight controls to effect an appropriate rate of climb.
3. Properly monitors air and helium pressure during climb.
4. Reduces rate of climb approaching pressure height and then identifies arriving at pressure height.
5. Monitors appropriate instruments, properly controls air and helium pressures, and ensures pressure height is not exceeded.
6. Descends from pressure height and follows proper procedure for managing pressure system.

B. TASK: IN-FLIGHT WEIGH-OFF

REFERENCE: Airship Flight Manual.

Objective. To determine that the applicant:

1. Exhibits commercial pilot knowledge by explaining the elements of an in-flight weigh-off.
2. Steers airship into the wind in level flight at an altitude of at least 500 feet (150 meters) AGL.
3. Reduces power to the specified airspeed and stabilizes airship.
4. Determines if the airship is being affected by updrafts or downdrafts.
5. Neutralizes elevator and rudder controls.
6. Observes attitude of the airship and pressure differential in the ballonets.
7. Determines trim and static condition.
8. Adjusts trim properly.

C. TASK: MANUAL PRESSURE CONTROL

REFERENCE: Airship Flight Manual.

Objective. To determine that the applicant:

1. Exhibits commercial pilot knowledge by explaining the elements of manual pressure control.
2. Controls the pressure manually as recommended by the manufacturer to a predetermined valve(s) setting.
3. Controls ballonet air balance by air valve operation.
4. Monitors operation of pressure valves and system.
5. Maintains a constant altitude, ±100 feet (30 meters).

D. TASK: STATIC AND DYNAMIC TRIM

REFERENCE: Airship Flight Manual.

Objective. To determine that the applicant:

1. Exhibits commercial pilot knowledge by explaining the elements of static and dynamic trim, to include the relationship between lift and load conditions, static trim, and center of buoyancy.
2. Establishes static trim for various weight conditions.
3. Establishes dynamic trim for various flight conditions.

IX. AREA OF OPERATION: NAVIGATION

A. TASK: PILOTAGE AND DEAD RECKONING

REFERENCES: AC 61-21, AC 61-23.

Objective. To determine that the applicant:

1. Exhibits commercial pilot knowledge by explaining the elements of pilotage and dead reckoning.
2. Follows preplanned course solely by visual reference to landmarks.
3. Identifies landmarks by relating surface features to chart symbols.
4. Navigates by means of precomputed headings, groundspeed, and elapsed time.
5. Makes a reasonable estimate of heading, groundspeed, arrival time, and fuel consumption to the destination.
6. Corrects for, and records, the differences between preflight fuel, groundspeed, and heading calculations and those determined en route.
7. Verifies airship's position within 3 nautical miles of flight planned route at all times.
8. Arrives at enroute checkpoints or destination within 5 minutes of ETA.
9. Maintains appropriate altitude, ±100 feet (30 meters) and established heading, ±20°.

B. TASK: DIVERSION

REFERENCES: AC 61-21, AC 61-23.

Objective. To determine that the applicant:

1. Exhibits commercial pilot knowledge by explaining the elements of diversion.
2. Selects an appropriate alternate airport and route.
3. Diverts promptly toward the alternate airport.
4. Makes a reasonable estimate of heading, groundspeed, arrival time, and fuel consumption to the alternate airport.
5. Maintains the appropriate altitude, ±100 feet (30 meters) and established heading, 20°.

C. TASK. LOST PROCEDURES

REFERENCES: AC 61-21, AC 61-23.

Objective. To determine that the applicant:

1. Exhibits commercial pilot knowledge of the elements of lost procedures.
2. Selects the best course of action when given a lost situation.
3. Maintains the original or an appropriate heading and climbs, if necessary.
4. Identifies the nearest concentration of prominent landmarks.
5. Uses navigation systems/facilities and/or contacts an appropriate ATC facility for assistance.

D. TASK. NAVIGATION SYSTEMS AND ATC RADAR SERVICES

REFERENCES: AC 61-21, AC 61-23; Navigation Equipment Operations Manuals.

Objective. To determine that the applicant:

1. Exhibits commercial pilot knowledge of the elements of navigation systems and ATC radar services.
2. Selects and identifies the appropriate facilities or coordinates.
3. Locates position of airship relative to the navigation facility or coordinates selected.
4. Intercepts and tracks a given radial or bearing.
5. Locates position using cross radials bearings, or coordinates.
6. Recognizes and describes station or waypoint passage.
7. Utilizes proper communication procedures when utilizing ATC radar services.
8. Maintains appropriate altitude, ±100 feet (30 meters).

X. AREA OF OPERATION: EMERGENCY OPERATIONS

A. TASK: ABORTED TAKEOFF

REFERENCE: Airship Flight Manual.

Objective. To determine that the applicant:

1. Exhibits commercial pilot knowledge by explaining the elements of an aborted takeoff.
2. Recognizes when dynamic lift is insufficient to continue takeoff.
3. Reduces power and applies reverse thrust (if applicable).
4. Maintains a straight course into the wind.
5. Establishes timely communication with ground crew.

B. TASK: ENGINE FAILURE DURING TAKEOFF

REFERENCE: Airship Flight Manual.

Objective. To determine that the applicant:

1. Exhibits commercial pilot knowledge by explaining the elements of engine failure during takeoff, to include loss of both engines under light and heavy conditions.
2. Promptly recognizes engine failure and utilizes prescribed emergency procedure, including use of checklist.
3. Maintains heading into the wind.
4. Manages air/helium pressures properly.
5. Follows checklist for engine restart or shutdown.
6. Returns for landing, as appropriate.

C. TASK: ENGINE FAILURE DURING FLIGHT

REFERENCE: Airship Flight Manual.

Objective. To determine that the applicant:

1. Exhibits commercial pilot knowledge by explaining the elements of engine failure during flight.
2. Attempts to determine cause for engine failure.
3. Uses checklist to attempt restart procedure.
4. Establishes near-equilibrium condition, if practical, when one engine fails.

5. Follows proper procedures for free-ballooning when both engines fail.

D. TASK: ENGINE FIRE DURING FLIGHT

REFERENCE: Airship Flight Manual.

Objective. To determine that the applicant:

1. Exhibits commercial pilot knowledge by explaining the elements of an engine fire during flight.
2. Attempts to extinguish fire in the affected engine using recommended procedures.
3. If fire persists, shuts down engine and uses appropriate checklist.
4. Prepares to land at earliest opportunity.

E. TASK: ENVELOPE EMERGENCIES

REFERENCE: Airship Flight Manual.

Objective. To determine that the applicant exhibits commercial pilot knowledge of the elements of envelope emergencies by explaining:

1. A puncture or rip in the gas envelope and/or in a ballonet.
2. Excessive helium loss in flight.
3. Rain/icing on envelope.
4. Emergency valve operations.
5. Emergency air-to-helium operations.
6. Recommended procedures to use when experiencing a specific envelope emergency.

F. TASK: FREE BALLOONING

REFERENCE: Airship Flight Manual.

Objective. To determine that the applicant:

1. Exhibits commercial pilot knowledge by explaining the elements of free ballooning.
2. Assesses airship static condition and determines ballast needs.
3. Establishes equilibrium in a timely manner.
4. Turns off all nonessential electrical equipment.
5. Determines cause of engine failure and attempts restart.
6. Selects suitable landing site and establishes communications with the crew.

7. Uses minimum helium valving and ballast dumping during descent.
8. Secures loose equipment.
9. Completes an appropriate emergency checklist.

G. TASK: DITCHING AND EMERGENCY LANDING

REFERENCE: Airship Flight Manual.

Objective. To determine that the applicant:

1. Exhibits commercial pilot knowledge by explaining the elements of ditching and emergency landing.
2. Simulates jettisoning ballast and fuel, considering potential fire hazard when jettisoning fuel.
3. Ensures airship is turned into the wind.
4. Instructs passengers in safety procedures, including use of life jackets if ditching.
5. Secures loose equipment.
6. Simulates securing all systems to minimize chance of fire or other damage.
7. Completes an appropriate emergency checklist.

H. TASK: SYSTEMS AND EQUIPMENT MALFUNCTIONS

REFERENCE: Airship Flight Manual.

NOTE: The examiner shall not simulate a system or equipment malfunction in a manner that may jeopardize safe flight or result in possible damage to the airship.

Objective. To determine that the applicant exhibits commercial pilot knowledge of the elements of systems and equipment malfunctions, appropriate to the airship used for the practical test, by explaining recommended pilot action for:

1. Control system/actuator malfunction.
2. Fuel starvation.
3. Electrical system malfunction.
4. Propeller malfunction.
5. Pressure system malfunctions.

XI. AREA OF OPERATION: POSTFLIGHT PROCEDURES

A. TASK: MASTING

REFERENCE: Airship Flight Manual.

Objective. To determine that the applicant:

1. Exhibits commercial pilot knowledge by explaining the elements of masting.
2. Maintains coordination with crew chief through use of proper hand signals and voice commands.
3. Remains in control of airspeed and positions airship properly.
4. Coordinates use of power and flight controls.
5. Places airship in proper trim and ballast when approaching the mast.
6. Completes an appropriate checklist.

B. TASK: POST-MASTING

REFERENCE: Airship Flight Manual.

Objective. To determine that the applicant:

1. Exhibits commercial pilot knowledge by explaining the elements of post-masting, appropriate to the airship used for the practical test.
2. Uses proper engine shutdown procedures.
3. Complies with equipment requirements for maintaining envelope pressure.
4. Ensures mast security relative to weather conditions.
5. Gives consideration to weather with airship on the mast.
6. Completes an appropriate checklist.

Made in United States
Troutdale, OR
11/03/2026

56844456R00049